Evandro Inocenti

Poultry farm automation using PLC

Evandro Inocenti

Poultry farm automation using PLC

Automation of temperature, lighting, feeding, hydration and ammonia gas concentration in a poultry house

ScienciaScripts

Imprint

Cover image: www.ingimage.com

This book is a translation from the original published under ISBN 978-613-9-69236-1.

Publisher:
Sciencia Scripts
is a trademark of
Dodo Books Indian Ocean Ltd. and OmniScriptum S.R.L publishing group

120 High Road, East Finchley, London, N2 9ED, United Kingdom
Str. Armeneasca 28/1, office 1, Chisinau MD-2012, Republic of Moldova, Europe
Printed at: see last page
ISBN: 978-620-8-13117-3

SUMMARY

This prototype describes the activities carried out during the final year of the Computer Engineering III course at the Universidade do Oeste de Santa Catarina (UNOESC). These activities include the conditioning of sensors and the development of routines for checking the poultry rearing environment. The routines include checking the concentration of ammonia gas, feed consumption and hydration, temperature monitoring, curtain control, and the supply of light during the night, as indicated by the veterinarian of the slaughterhouse associated with the poultry farmer, using a programmable logic controller. In this way, it is possible to monitor the birds' development more precisely and better control the dosage of nutrients they consume. Checking the accumulation of ammonia gas, monitoring the temperature and controlling the curtains is necessary to ensure that the birds develop uniformly, avoiding disease and a high mortality rate. Programming routines to carry out this management makes the system autonomous, reducing human contact with the birds. By calibrating the sensors and conditioning the routines to check the environmental conditions, it is possible to check for gas accumulation and keep the ambient temperature close to the ideal for rearing the birds by providing methods for renewing the ambient air so that the birds can develop healthily.

Keywords: Automation, PLC, Poultry farming.

SUMMARY

CHAPTER 1

INTRODUCTION

In recent decades, Brazilian poultry farming has shown high growth rates and is one of the agricultural activities that has developed the most, conquering the most demanding markets and making Brazil the world's third largest producer and leading exporter, according to the Ministry of Agriculture (MAPA, between 2008 and 2016). This may be due to the efficient capital turnover provided, since the birds' rearing time, from housing to slaughter, is between 43 and 50 days. However, along with the economic gains resulting from the growing intensification of this branch of activity, there are the concerns arising from the need to constantly promote animal welfare at all stages of rearing (AVES, 2016).

1.1 DESCRIPTION OF THE PROBLEM

Small adjustments and care taken with the internal aspects of the poultry rearing environment are important for animal welfare, as they guarantee quality in the rearing and prevent the birds from developing diseases. In order to ensure that the environment is always close to ideal, constant monitoring is required in poultry houses, resulting in a great deal of labour. In the absence of this, events can occur that jeopardise the birds' comfort, such as changes in the ambient temperature recommended for each week of the rearing period, lack of water and feed and low light and/or lack of air circulation, causing harmful gases to accumulate. All these factors can cause an imbalance in the rearing process, directly influencing the chicken's development (COOB-VANTRESS, 2012a).

1.2 BACKGROUND

Carrying out activities such as checking water, temperature and controlling curtains requires constant intervention on the part of the poultry farmer. Therefore, automation is an alternative to help develop the necessary care in the rearing environment, as well as being effective and offering greater precision, offering 24-hour monitoring and with activities being carried out according to local needs, reducing the need for labour in the poultry house, reducing costs and increasing productivity.

Thus, the use of an automation system with a Programmable Logic Controller (PLC) improves the quality and safety of the service, as well as ensuring that the environment in the aviary provides for the birds' well-being, and that corrective action is taken automatically for each event that alters the ideal parameters.

In 2015, the state of Santa Catarina became the 2° largest chicken exporter in the country, slaughtering more than 1 billion birds a year. This sector has more than 10,000 poultry farmers and employs 40,000 people directly (Gl, 2015).

As this is a growing sector in the state, the use of automation could facilitate and improve the quality of poultry farming. In addition, an automation system can help the birds to develop uniformly, and can provide greater quality to the environment through uninterrupted control, reducing the possibility of bird diseases and, consequently, increasing productivity.

1.3 OBJECTIVES

To develop an automation system for poultry houses using a PLC to control temperature, lighting and ammonia concentration, and to monitor water and feed consumption during the poultry rearing period in a traditional, semi-automatic poultry house.

1.3.1 Specific Objectives for Course Conclusion Work III

- Integrate curricular internship projects I and II into the Course Conclusion Work;
- Calibrate the developed algorithm;
- Develop sensor boards;
- Integrating sensors with the PLC;
- Developing a supervisory system;
- Demonstrate how the prototype works;

CHAPTER 2

AUTOMATION OF THE POULTRY REARING ENVIRONMENT

This chapter will briefly describe the poultry rearing environment, focusing on preventing the build-up of ammonia gas, hydration and feeding the birds.

2.1 FORMATION OF AMMONIA

Ammonia gas is one of the pollutants found in high concentrations inside poultry houses. This is mainly due to the fact that broilers are kept indoors most of the time (OWADA et al., 2007). This concentration of ammonia in high proportions can cause damage to the birds' respiratory system, such as coughing, a reduction in growth rate, an increase in the number of bacteria in the lungs and in the mortality rate of the chickens, even causing possible blindness (FERREIRA, 2010).

In Brazil, there are no legal limitations on the exposure of poultry to ammonia, however, the import market for poultry meat stipulates that the degree of exposure of poultry should not exceed 20 ppm (parts per million). However, in the last few weeks of production, ammonia can reach values of up to 50 ppm if the poultry house is not well ventilated (OLIVEIRA; MONTEIRO, 2013).

According to a study by Ferreira (2010), ammonia also affects human health. For farm workers, the symptoms caused by the gas are similar to those affecting poultry, such as coughing, difficulty breathing, fatigue, fever, among others. Ammonia becomes noticeable to the farmer after reaching a limit of 50ppm. In Brazil, legislation allows a worker working an 8-hour shift to remain in environments with a maximum concentration of this gas of up to 20 ppm (ZANATA, 2007).

Factors such as temperature, relative humidity and ventilation of the internal environment have a direct influence on the accumulation of ammonia. A low ventilation rate makes the broiler house more humid, thus favouring a higher concentration of ammonia in the air (MIRAGLIOTTA, 2005). The poultry litter is the place where the birds stay during their entire rearing period. It is laid out on the ground and can be made up of materials such as pine wood shavings, shredded grass, corn cobs and others (BASSI et al., 2006).

The use of fans inside the poultry house allows air to move around, thus renewing the environment. Care with ventilation is not only necessary during periods of high temperatures, but throughout the birds' rearing period, in order to avoid the accumulation of ammonia (ZANATA, 2007).

1.1 HYDRATION

Water is an essential nutrient that influences practically all the physiological functions of birds. A bird's body is made up of around 75 per cent water (COOB-VANTRESS, 2012a). A loss of 10 per cent of weight due to dehydration will cause a drop in development, while a loss of around 20 per cent of body water can lead to death (AVILA et al., 1992).

The chickens must have access to clean, potable and fresh water, except when there is a need to provide medication (CARE, 2014). In these cases, the veterinarian of the associated slaughterhouse will advise.

Low feed consumption by birds can be directly related to insufficient water consumption. A bird drinks an average of 2 to 3 litres of water for every kg of feed consumed. Water consumption varies according to age, temperature and the type of feed provided, as this varies according to the birds' development cycle (AVILA et al., 1992).

Even if the water is potentially potable for human consumption, it still has to pass laboratory tests. For poultry, chlorine treatment is necessary, as it acts directly on bacteria, viruses and other organic materials that render the water harmless from a microbiological point of view. Chlorine is most effective when used in water with a pH of 6.0 to 7.0. This level results in a higher percentage of hypochlorous acid ions, which have a strong disinfectant action (COOB-VANTRESS, 2012a).

Table 1 shows the ideal amount of water to consume daily for 1,000 chickens.

Table 1 - Average daily water consumption per 1,000 chickens

Weeks	1	2	3	4	5	6	7	8	9	10
Litres/day	38	57	76	99	129	160	186	208	227	246

Source: (AVILA et al., 1992)

2.3 FOOD

Chickens should be fed a healthy diet, i.e. taking into account their age, stage of production and type (male, female or mixed) (CARE, 2014). The nutritional diet for broilers is designed to offer nutrients, vitamins and minerals. These components need to act together for the correct skeletal and muscular development of the chicken (COOB-VANTRESS, 2012a).

In order for the chicken to maintain uniformity and remain close to the body mass target in a diet, its feeding and management are fundamental to maintaining quality during its breeding life (ROSS, 2013). In the first 12 hours of the birds' lives, the feed should be laid out on paper liners evenly distributed in the space where they are placed after being housed, thus facilitating access for feeding (BASSI et al., 2006). The feed should be offered freely to the birds, and for the first 7 days there should be trays available for easy access to the food (BIGSAL, between 2006 and 2016).

In order for the producer to effectively monitor the development of the birds, Coob-Vantress (2012b) provides tables for monitoring their daily weight, from the time they arrive at the poultry house until they leave for slaughter. These tables are standardised for each species: male, female or mixed. The tables provided below have been adapted by the author to show the weights weekly up to the 42nd day of rearing, after which they will be shown daily up to the 49th day, which is the range used by the slaughterhouses in the Midwest region of Santa Catarina for slaughter, the values shown are the originals from the manual (COOB-VANTRESS, 2012b).

Table 2 shows the weekly performance targets for mixed-type birds.

Table 2 - Performance Targets for Mixed Chickens

Age (days)	Weight	Daily Gain(g)
0	42	-
7	177	29
14	459	49
21	891	70
28	1436	83
35	2067	94
42	2732	95
43	2826	94
44	2919	93
45	3011	92
46	3192	91
47	3281	90
48	3369	89
49	3369	88

Source: Adapted from (COOB-VANTRESS, 2012b)

Table 3 shows the weekly weight targets for female birds.

Table 3 - Performance Targets for Female Chickens

Age (days)	Weight	Daily Gain(g)
0	41	-
7	175	29
14	443	47
21	844	66
28	1341	77
35	1914	85
42	2511	86
43	2596	85
44	2679	83
45	2760	81
46	2841	81
47	2922	81
48	3003	81
49	3084	81

Source: Adapted from (COOB-VANTRESS, 2012b)

Table 4 shows the weight and daily gain that male birds must achieve in order to be close to proper development.

Table 4 - Performance Targets for Male Chickens

Age (days)	Weight	Daily Gain(g)
0	43	-
7	179	29
14	475	51
21	938	74
28	1531	92
35	2217	101
42	2953	109
43	3060	107
44	3165	105
45	3268	103
46	3369	101
47	3468	99
48	3565	97
49	3660	95

Source: Adapted from (COOB-VANTRESS, 2012b)

As can be seen in both tables, from the 42nd day of production onwards, the chickens tend to decrease their daily gain, but maintain their development rate. This may be due to the nutrition that makes up the feed for this period.

In semi-automatic poultry houses, the feed is supplied automatically via limit switches installed in the boxes that store the feed. When these are deactivated, they switch on the motor, which has a spring attached to it and passes through a pipe, causing the feed to move from the silo to the boxes as the motor turns, overlapping the sensors. This method of operation is illustrated in Figure 1.

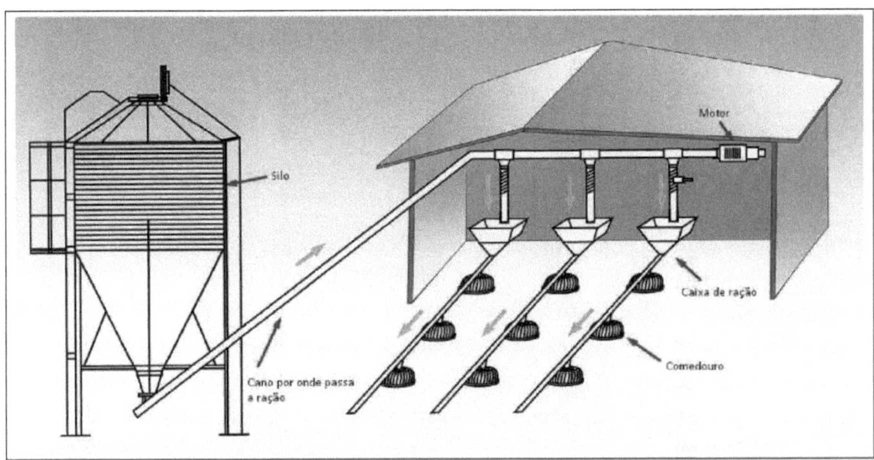

Figure 1 - Feed supply in semi-automatic aviaries

Source: Adapted from (5PIGR0UP, between 2010 and 2017)

2.4 RELATED RESEARCH

This section cites some of the related works that contributed to its development.

- AUTOMATION OF A GRINDING PROCESS FOR ANIMAL FEED.

Academic: Rodrigo Reisdorfer.

Institution: UNOESC.

In this prototype, academic Reisdorfer (2011) describes the automation process for milling in an animal feed industry, the aim of which is to automate a mill

that was being built at the time. This work was of great importance in defining the sensor for feed control, given that the silo where the academic's study was carried out is similar to the one used in this prototype.

- SYSTEM FOR MONITORING AMMONIA GAS IN POULTRY HOUSES.

Academic: Jean Luiz Zanatta.

Institution: UNOESC.

The prototype developed by Zanatta (2014) consists of developing a web application to display data obtained by an ammonia sensor, which is part of a piece of *hardware* he developed. The work in question played an important role in defining part of the rationale and understanding how the ammonia sensor was used to acquire signals relating to the identification and concentration of this gas.

- REMOVAL OF AMMONIA GENERATED ON POULTRY FARMS AND ITS USE IN FUEL CELLS AND AS FERTILISER.

Academic: João Moutinho Ferreira.

Institution: INSTITUTO DE PESQUISAS ENERGÉTICAS NUCLEARES.

In this prototype, Ferreira (2010) emphasises how harmful ammonia gas can be for birds and for the people who work in this environment. The author used a composite absorber with chemical products to retain the ammonia. The work carried out by Ferreira (2010) played a major role in the development of the prototype presented in this report, when he mentions ammonia gas, especially when he talks about how harmful this gas is to health.

CHAPTER 3

THEORETICAL BACKGROUND

This section describes the *hardware* and *software* technologies that will be used to develop the complete prototype.

3.1 PROGRAMMABLE LOGIC CONTROLLER

The Programmable Logic Controller (PLC) is defined as an industrial computer capable of storing instructions for implementing control functions, performing logical and arithmetic operations, manipulating data and communicating via a network, and is used to control automated systems (GEORGINI, 2012).

Their structure is designed to withstand adverse situations, such as: vibrations, temperature variations, electrical noise, small voltage variations, among others (PAREDE; GOMES, 2011). The vast majority of PLCs are available in modules. As such, they can be configured by the user, who adds them according to the needs of the task to be performed. There are also compact forms in which it is already configured and the user cannot make any changes (PAREDE; GOMES, 2011).

Figure 2 shows the image of a common Programmable Logic Controller installed on a bench, currently available on the market. This image is made up of the following components:

1- Power supply;

2- Electrical connections between I/O devices and the PLC;

3- Input and output devices and pushbuttons;

4- Expansion devices (network, serial);

5- Programmable Logic Controller;

6- PWM controller;

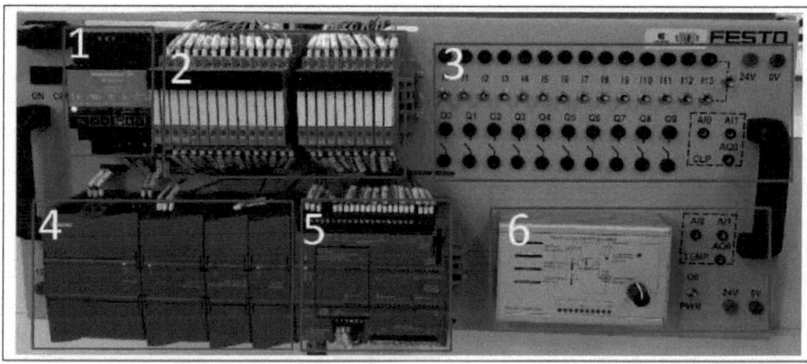

Figure 2 - Siemens Simatic S7-1200 PLC
Source: The author

The structure of a PLC consists of a CPU, a memory block, I/O modules, a power supply and a base or *rack*. The CPU can be a microcomputer or microcontroller. It controls and supervises all the operations carried out on the PLC's electronic circuits by means of instructions stored in memory (PAREDE; GOMES, 2011).

The memory block is where the programmes, manufacturer *firmware*, code and data from the algorithm developed by the user are stored. The PLC's memory is organised hierarchically. The top module is next to the processor, then come the registers, then the cache memory, main memory (RAM) and, continuing, the memories that store the manufacturer's programmes, ROM, EPROM, EEPROM and Flash (PAREDE; GOMES, 2011).

I/O modules connect the interface to the external system. The input modules receive signals from voltage or current electrical quantities, which can be from sensors or field pushbuttons. The output modules communicate the action to be taken to the actuators, depending on the configuration of the *software* developed to handle the input signals (PAREDE; GOMES, 2011). For the PLC model in question, the electrical signals to the PLC can be digital with DC voltage of 12V or alternating voltage of 110V, or analogue with current variations between 4 and 20mA, or DC voltage between 0 and 10V (GEORGINI, 2012).

Communication between the CPU, the memory module and the I/O devices is done via internal buses under the control of the CPU. The speed of PLC operations varies according to the *clock* frequency (PAREDE; GOMES, 2011).

The power supply is responsible for supplying power to the CPU and I/O circuits (GEORGINI, 2012). Power supplies can be internal or external. Internal power supplies can be *slotted* or mounted in the PLC case. These supplies generally provide 24V signals and low voltages, and are intended for sensors and actuators. If the module requires more power, the designer includes an external supply to fulfil this need (PAREDE; GOMES, 2011).

The *rack* provides the connection between the CPU, I/O modules and the power supply. It contains the communication bus between the two and contains the data, address, control and supply voltage signals (GEORGINI, 2012). Diagram 1 shows how a basic PLC structure is laid out using the blocks described.

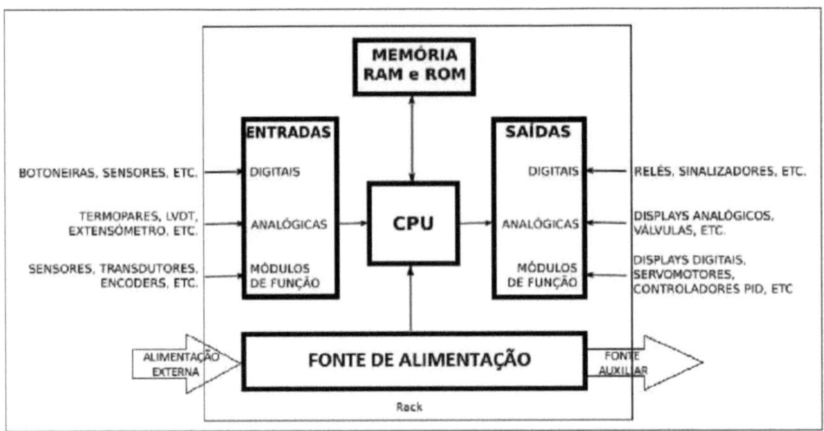

Diagram 1 - Basic PLC structure

Source: Adapted from (GEORGINI, 2012)

In the structure of the PLC in question, the CPU is centralised, where it receives power from the power supply, and the data obtained by the sensors that will be connected to its inputs, the CPU executes the instructions by communicating with the memory, and generates the actions for the outputs. These actions can be motor activation or power supply, for example.

3.2TIA PORTAL

TIA Portal, which stands for Totally Integrated Automation Portal, is an integrated platform for implementing automation solutions, allowing all automation processes to be designed in the best way, from a computer screen

(SIEMENS, between 2005 and 2016). This system architecture is the foundation for open connectivity and maximum interoperability across multiple devices to transform them into a fully integrated automation system (SIEMENS, between 2000 and 2015). Figure 3 shows the production life cycle with the TIA Portal structure. The TIA Portal platform comprises HMI programming, programming technologies, driver configurations, PLC programming, network configuration and online diagnostics. Generally speaking, with this tool it is possible to configure the drivers and the network for communication between a microcomputer and the PLC, as well as the possibility of choosing from a number of programming languages, which will be mentioned later, to programme the algorithm for which it will be used.

STEP 7 *Professional software* is a programming tool for SIMATIC automation systems. This *software* is included in the TIA Portal package. Using this tool, you can utilise a number of functions to automate a system, such as:

-*Hardware* configuration and parameterisation[1] ,

-Establishing a connection;

-Programming ;

-Documentation ;

The parameterisation of these functions can be found in Siemens (between 2005 and 2015).

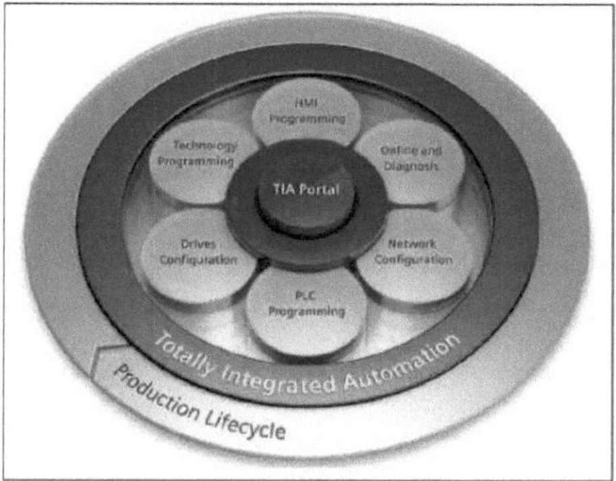

Figure 3 - TIA Portal

Source: (SIEMENS, between 2000 and 2015)

3.3 PROGRAMMING LANGUAGE

This section introduces the programming language used to design PLC *software*.

3.3.1Ladder *Diagram*

The *Ladder Diagram* programming language, or simply Ladder, as it is known, was the first to be created for use in PLCs. The determining factor for the language's acceptance in PLCs was that it had a graphical interface based on symbols similar to those found in electrical circuits, such as contact and coil symbols (GEOR-GINI, 2012). This language is made up of graphic signals and is suitable for programmers with knowledge of electromechanical logic (PRUDENTE, 2013). Figure 4 shows a simple example of Ladder programming.

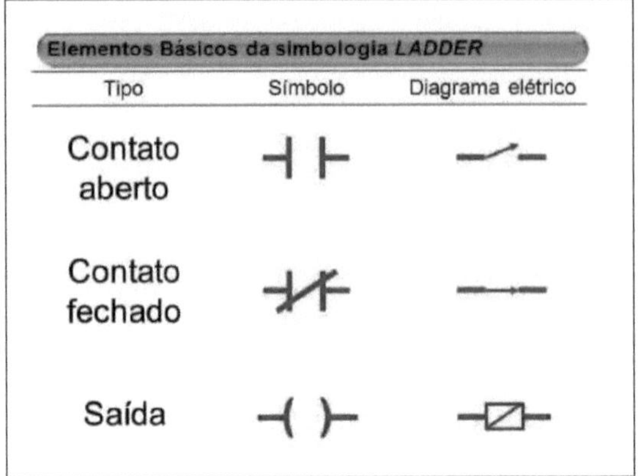

Figure 4 - Basic Elements of Ladder Programming

Source: (WORDPRESS, between 2005 and 2014)

The Ladder diagram is made up of the power bar, which controls all the *input* elements, the common return or mass, which connects all the *output* variables, the test zone, where the process logic is designed, and the action zone, which is used for the output variables (PRUDENTE, 2013). See Figure 5 for an example of the structure of a simple piece of *software*. In this code, the output will be active when inputs A, B and C are on, or when inputs A and C are on and D is off.

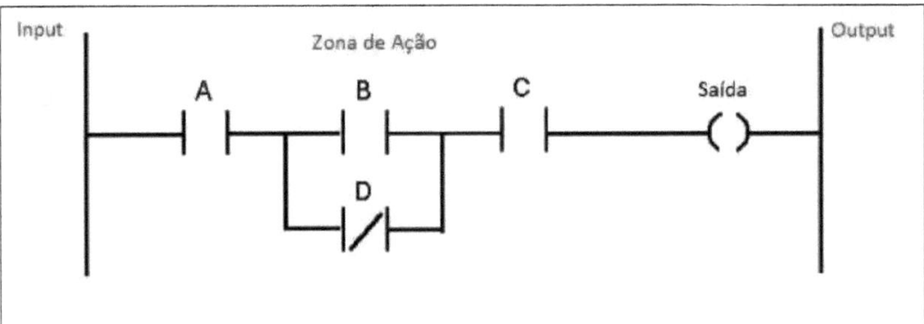

Figure 5 - Example of Ladder programming
Source: Adapted from (SILVEIRA, between 2005 and 2018)

3. 4SENSORS

Sensors are devices that are sensitive to some form of energy from the environment, which can be light, thermal or kinetic, relating information about quantities that need to be measured, such as temperature, speed, current, etc. (THOMAZINI; ALBUQUERQUE, 2012). Sensors are classified in two ways: analogue and digital. Analogues can take on any values in the output signal over time, as long as they are within their operating range. Physical quantities such as temperature, flow, luminosity, humidity and pressure can take on any value over time (THOMAZINI; ALBUQUERQUE, 2012). Digital sensors, on the other hand, can only take on two values in their signal over time, which can be interpreted as zero or one. There are no physical quantities with these values, but they are thus displayed to the control system after being converted by the transducer's electronic circuitry (THOMAZINI; ALBUQUERQUE, 2012).

3.4.1 Temperature transducer - Pt 100

A transducer is the name given to a complete device made up of a sensor and interface circuits. The transducer transforms a physical quantity, such as temperature, into a voltage or current signal that can be interpreted by a control system (THOMAZINI; ALBUQUERQUE, 2012).

The Pt 100 temperature transducer works by changing the electrical resistance of the element as the temperature varies. The temperature coefficient, which is the variation of resistance with temperature, is specified with the average variation between 0 and 100°C, which is 0.385 per °C. Stability remains constant, both in the

and the flat film sensors have a platinum film (NOVUS, between 2010 and 2016). Figure 6 illustrates a model of the PtlOO temperature sensor available on the market, where demarcation 1 shows the sensor that captures the temperature and demarcation 2 shows the power supply and output to connect to the PLC.

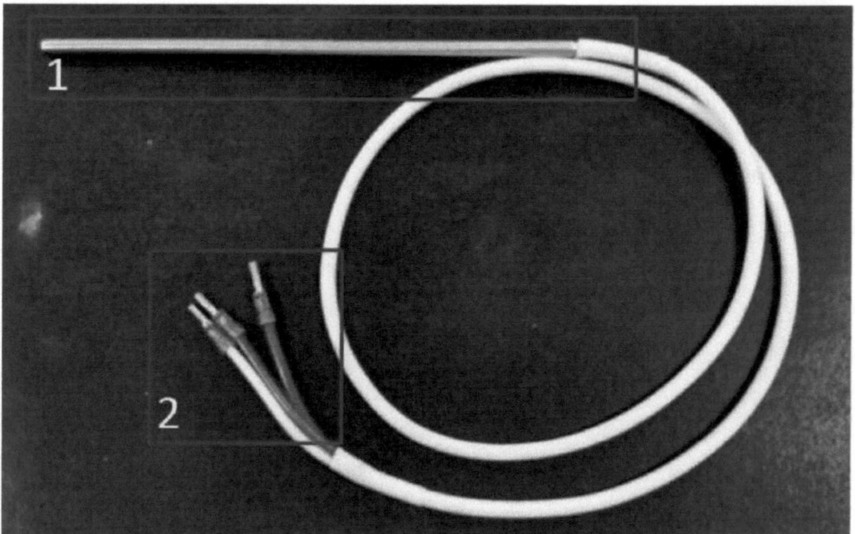

Figure 6 - PtlOO sensor

Source: (SIRICHOTE, between 2010 and 2016)

3.4.2 Water Level Sensor

Level sensors make it possible to detect whether a certain level has been reached or not. High level detection for filling and low level detection for extracting the monitored product. The water level sensor works like a magnetic switch, shorting out when moved upwards, thus activating a pump or carrying out a specific task (USINAINFO, between 2010 and 2016). Figure 7 shows the water level sensor with float in its two physical states. Image *a* shows the level sensor with the logic state closed, while image *b* shows the level sensor open.

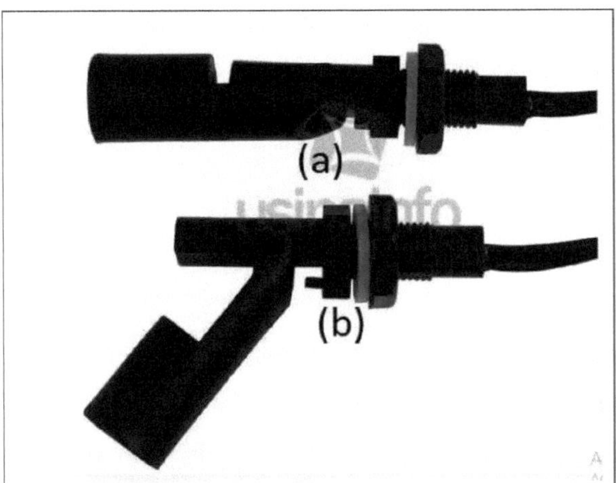

Figure 7 - Water level sensor with float
Source: (USINAINFO, between 2010 and 2016)

3.4. 3Inductive sensor

Inductive sensors are electronic devices capable of detecting metallic objects when they are close to the sensor. This detection occurs without the need for contact between the metal and the sensor (SENSE SENSORES E INSTRUMENTOS, 2002). The operation of the sensor is based on an electromagnetic field that is generated by the oscillator and the coil arranged intemally, this electromagnetic field is at the end of the device. As a metal is approached, there is a decrease in the energy supplied to the electromagnetic field, so when this signal becomes too low, the trigger circuit realises the change and alters the output voltage, responding logically at a high or low level (SILVEIRA, 2016). Figure 8 below shows a type of inductive sensor.

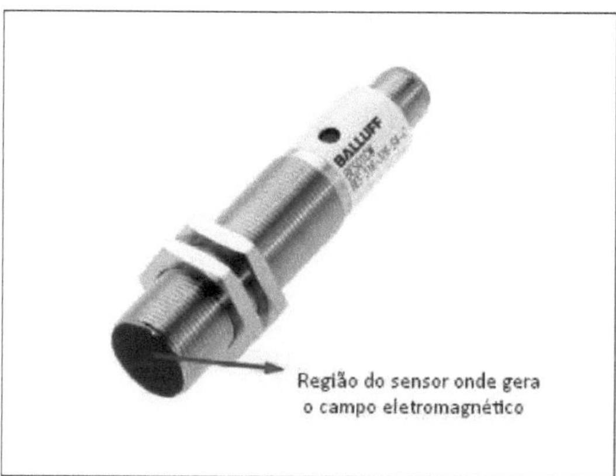

Figure 8 - Inductive sensor
Source: Adapted from (SILVEIRA, 2016)

3.4. 4Flow Sensor

Flow sensors are used to check fluids. This type of sensor works with a turbine, consisting of an internal propeller, a magnet and a Hall-type sensor. When the fluid passes through the propeller, it rotates causing the magnet attached to it to pass through the sensor, thus detecting the number of turns made (PORTA, 2016). Figure 9 shows an example of a flow sensor.

Figure 9 - Flow sensor

Source: (INSTITUTE, between 2010 and 2017)

3.4. 5Ammonia Gas Sensor

The MQ-135 gas sensor consists of a carbon dioxide semiconductor, which is a transparent conductive oxide and has lower conductivity in air. This sensor, placed in an atmosphere where there are polluting gases, causes its conductivity to increase with the gas concentration. This sensor is good at detecting ammonia, sulphide vapour and benzene (WAVESHARE, Between 2000 and 2012). Figure 10 shows the gas sensor described.

Figura 10 - Gas sensor

Source: (SILÍCIO, between 2010 and 2017)

3.4.6 LDR Light Sensor

The LDR light sensor is a component whose resistance varies according to the intensity of the light shining on it. The more light that falls on the sensor, the lower the resistance (SUNROM TECHNOLOGIES, 2008). Figure 11 below illustrates the LDR light sensor.

Figura 11 - LDR sensor

Source: (MOTA, 2015)

3.5 SYSML

The SysML graphical modelling language is a tool that brings together a subcon with UML 2 and provides additional extensions to meet the requirements of Systems Engineering, which include processes, people, information, *firmware, software* and *hardware.* This language is widely used in systems modelling, supporting the analysis, specification, prototyping and verification of complex systems (OMG, OBJECT MANAGEMENT GROUP, 2010).

The diagram below shows a SysML model that shows how UML diagrams have been reused, as well as the new SysML diagrams.

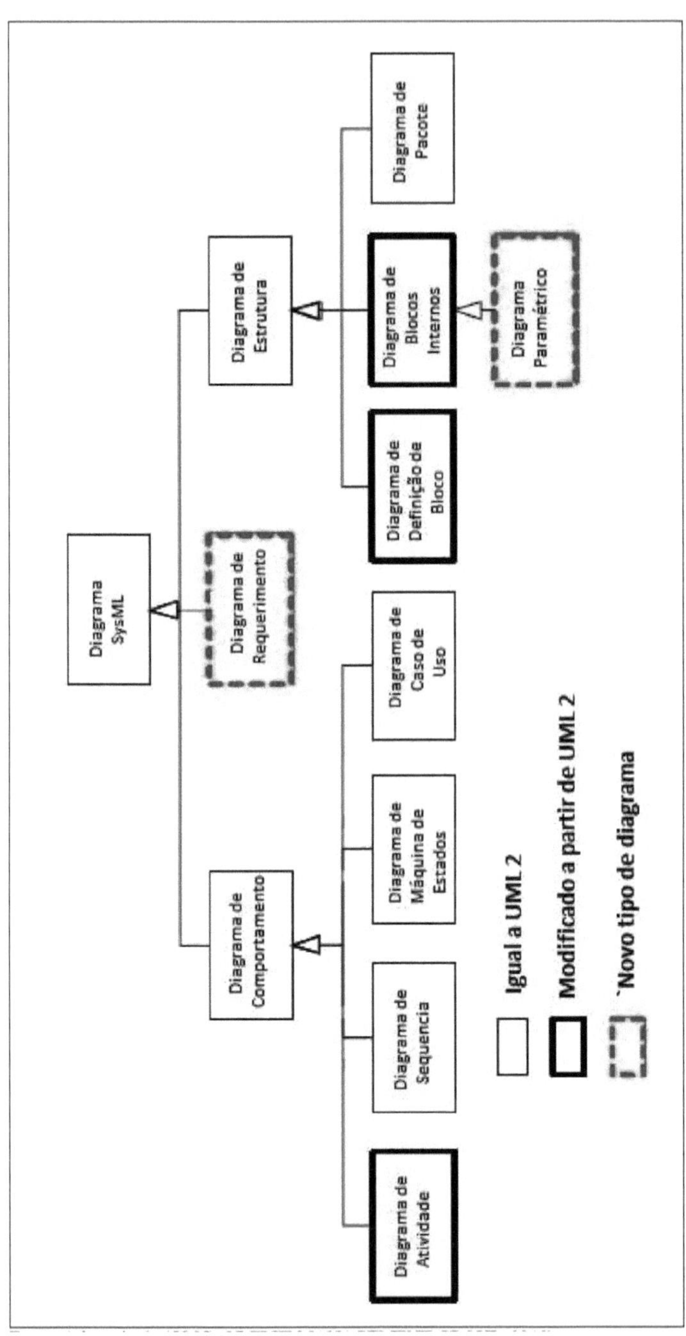

Diagram 2 - SysML model

CHAPTER 4

METHODOLOGY

The research methodology adopted for the prototype is experimental (PÁDUA, 2004). This consists of developing an application and applying it to correct the problem mentioned, checking for other research in the area of study and identifying theoretical bases with references in order to be able to recognise management variables and techniques. The prototype was developed with the aim of being autonomous.

This eliminates the need for human intervention when it comes to controlling curtains, activating/deactivating fans and managing the lighting time for periods with little light, such as at night.

The study carried out, based on books and articles in the field of poultry breeding, showed that during the growth process of poultry, climate control is essential so that the birds can develop healthily. An application was therefore developed to monitor and control environmental variables through ventilation, lighting, water, feed and ammonia control routines, thus offering a stable environment that provides animal welfare during the rearing period.

For the development of TCCIII, bibliographical research was carried out on the sensor technologies that could be used for the general development of the prototype. Among these sensors are the induction sensor to control feed consumption; the flow sensor, to monitor the amount of water drunk by the birds; and the ammonia sensor, to identify the concentration of this gas present in the environment, and when it is higher than acceptable, provide methods for reducing it. We also searched for bibliographic material on the ideal conditions for the birds' rearing environment to provide them with comfort.

4.1 COMPUTER SYSTEM OVERVIEW

The PLC operates in cycles, starting by loading the variables into *the firmware in* memory, then scanning the input ports, executing the programmed routines, updating the outputs and returning to check the input ports. This process can be seen in Flowchart 1.

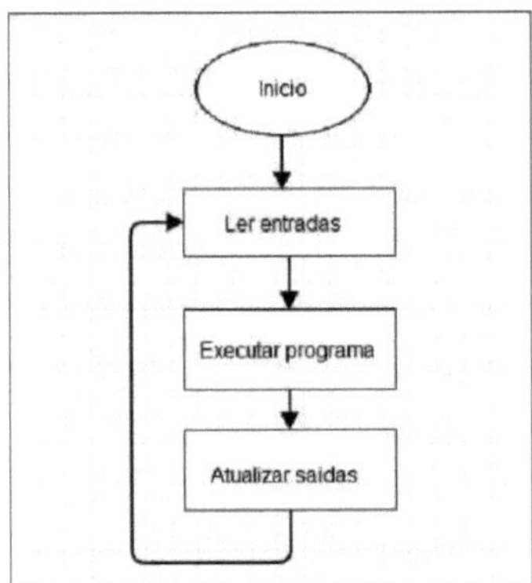

Flowchart 1 - PLC Operating Flowchart

Source: the author.

The ***software*** was programmed in the Ladder language. The ***software*** was developed using the tool provided by the PLC manufacturer itself, which in this case is Siemens. This tool is designed for use with Siemens PLCs and is part of the TIA Portal package.

The proposed automation systems are arranged in different routines, which carry out specific tasks for each case of ambience. These routines can be seen in Diagram 3. This process consists of constantly checking the ammonia, temperature and light sensors, thus carrying out subroutines such as checking the water tank, the light routine, the ammonia routine and the ventilation routine.

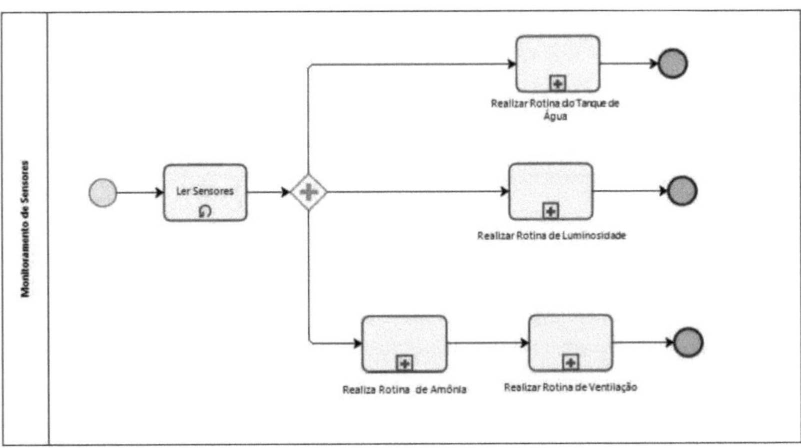

Diagram 3 - Main Process Diagram
Source: The author.

Ammonia is managed and controlled using the values acquired from the environment by the sensor. It will work as follows: When an ammonia concentration above 20 ppm is detected and the poultry house curtains are closed, both will be partially opened and the fans will be activated for a predetermined time of 3 minutes to circulate and renew the ambient air; if the curtains are open, only the fans will be activated. This process of checking the ammonia concentration will take place every 20 minutes so that there is not too much loss of temperature in the room, especially on colder days. Process diagram 4 shows how the routine described above will be developed.

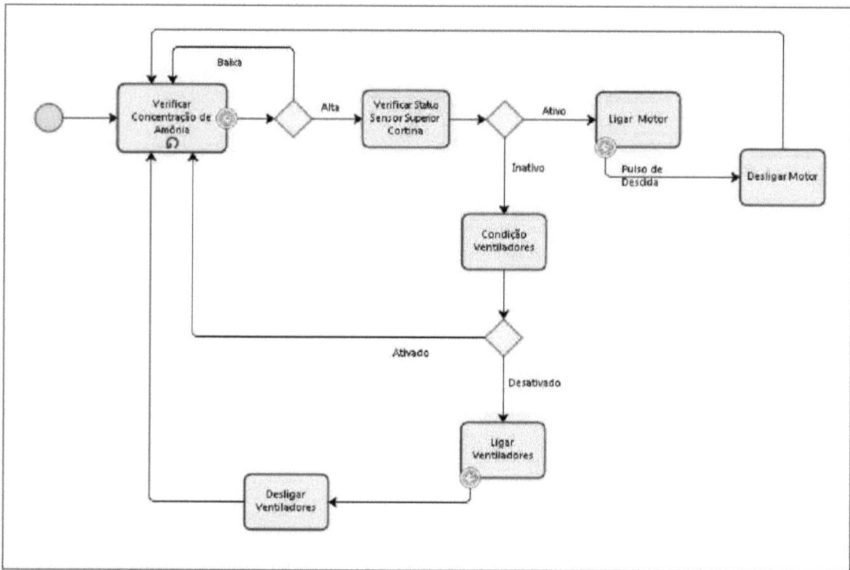

Diagram 4 - Ammonia Check Diagram
Source: The author.

The system was developed to identify the accumulation of ammonia, manage the water tank, the supply of electric light during the night, and the consumption of feed and water so that it is possible to control how much of these nutrients are being consumed by the birds.

Diagram 5 illustrates the classes that make up all the desired stages for the final prototype, including a main method and subclasses, each of which has its own verification and control variables for the management routine to which it belongs.

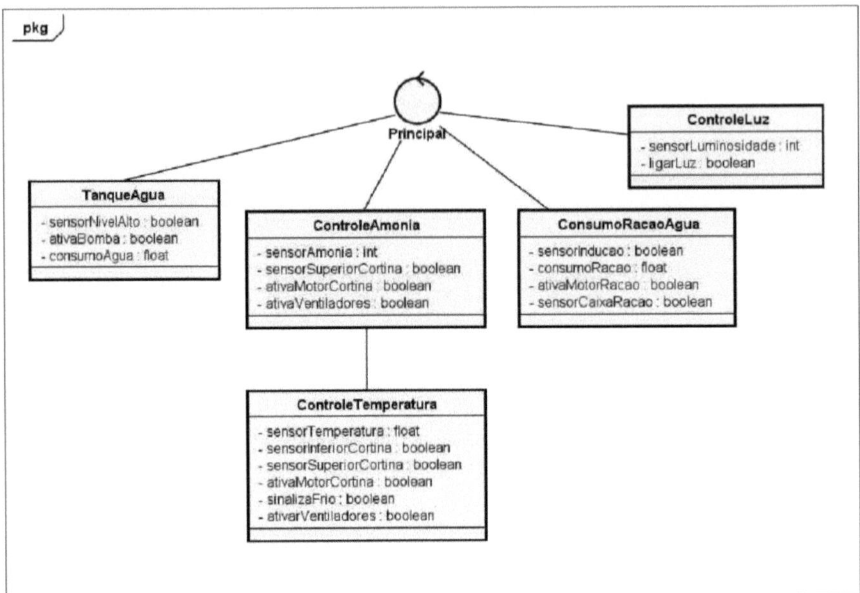

Diagram 5 - ***Software*** Conceptual Model
Source: The author.

4.2 EMBEDDED SYSTEM DEVELOPMENT

The embedded system was developed and designed on isolated circuit boards, thus allowing adjustments to be made to the circuit if necessary, preventing electrical noise and poor contact from causing the sensors to fail to communicate. Once the boards were made, the sensors were calibrated and integrated into the PLC so that the information they captured could be displayed via the HMI and used to constantly check the birds' rearing environment.

When developing and calibrating the water flow sensor, it was necessary to amplify the sensor's output signal so that it exceeded 10 V, which is the minimum signal required for the S7-1200 PLC's digital input. To make this possible, an optocoupler was added, powered by 24 V, so that when the sensor detected the passage of water, it would send a voltage pulse to the PLC, allowing it to count the turns made by the valve. The circuit developed for this signal conditioning is shown in Figure 12 below.

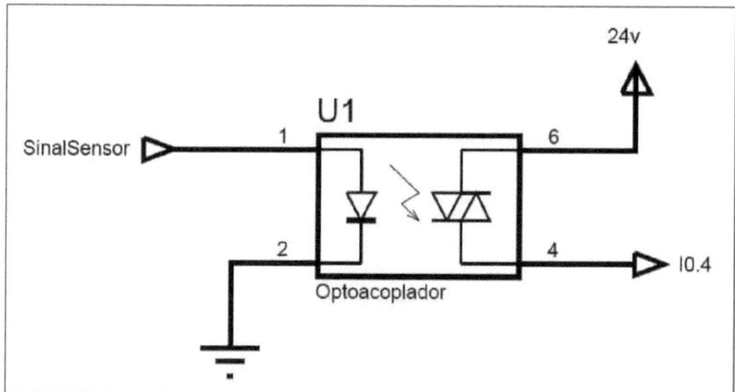

Figure 12 - Valve signal amplification circuit
Source: The author

Figure 13 shows the front view of the board on which the optocoupler is arranged along with the interconnections for the correct operation of the water flow sensor.

Figure 13 - Valve signal amplification circuit - front

Source: The author

Figure 14 shows a view of the back of the water flow sensor board.

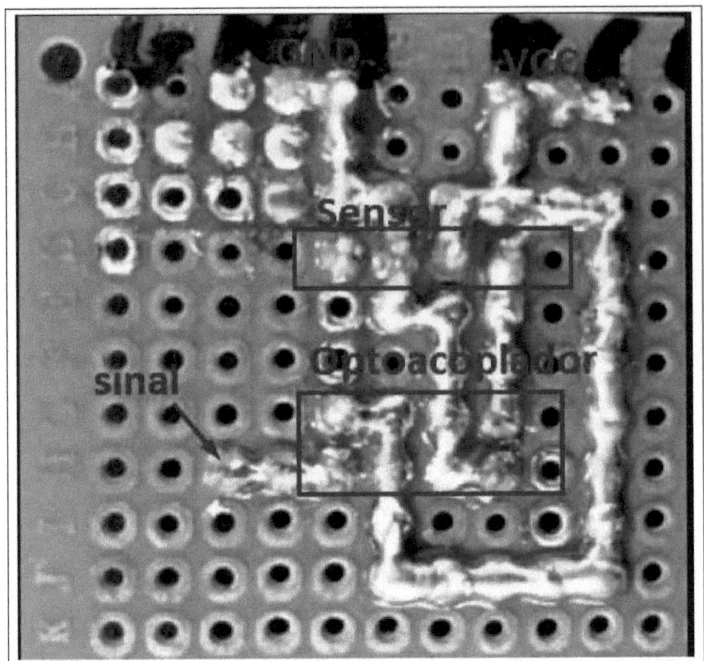

Figure 14 - Valve signal amplification circuit - reverse side

Source: The author

For the luminosity sensor, which is used to check the luminosity of the environment and activate the electric light according to the programme, a circuit was developed to activate the digital input 10.3 of the clp, thus following the routine implemented for supplying electric light at night. The luminosity sensor (LDR) is powered by a voltage of 5 V, changing its resistance according to the incidence of light on it. Using a 10 kQ resistor and a BC546 transistor, a voltage was configured to activate a relay that allows energy to pass to the clp's digital input, thus executing the routine for supplying electric light. Figure 15 shows this electrical circuit.

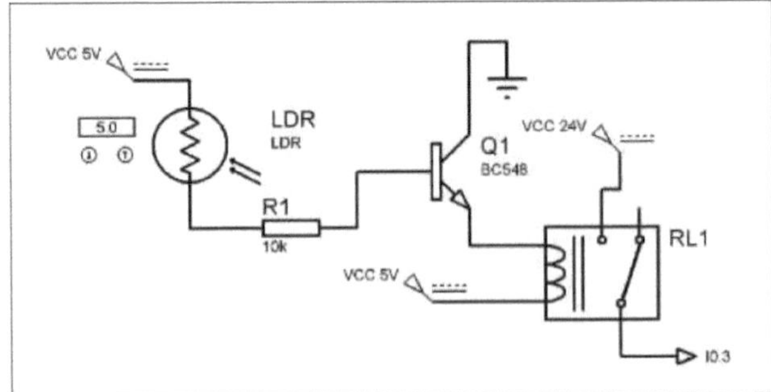

Figure 15 - Light sensor electrical circuit
Source: The author

With the circuit duly assembled and tested, the board was made, which can be seen in Figure 16.

Figure 16 - Light sensor circuit board

Source: The author

Figure 17 shows the connection of the gas sensor, which is supplied with 5 V, and the output signal connected to the PLC's IW64 analogue port.

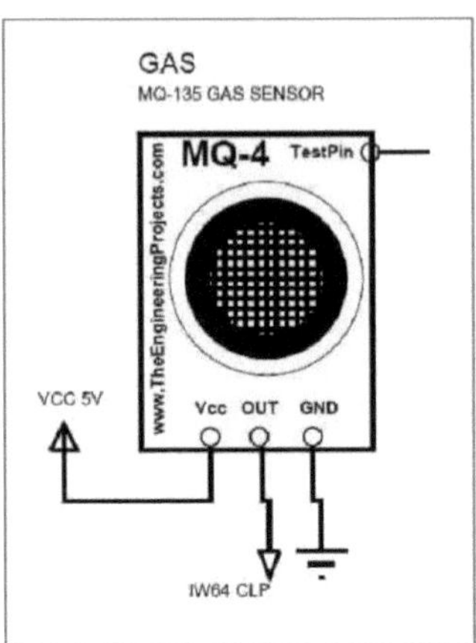

Figura 17 - Gas sensor connection
Source: The author

Figure 18 shows the gas sensor connected to the board designed to integrate with the PLC.

Figure 18 - Gas sensor board

Source: The author

Diagram 6 shows the organisation of *the* prototype, where you can see the *firmware* in which the management routines were programmed; the sensors responsible for obtaining data from the environment (positioned on the left); the actuators, which carry out corrective tasks (positioned on the right) to keep the environment stable; and the supervisory system, which makes it possible to observe the data obtained by the sensors, and the corrective actions being applied when necessary. It should be emphasised that at this initial stage the poultry farmer will only have interaction with the *hardware* to start rearing the birds when they are housed, and end when they leave for slaughter, which in this case is a start switch connected to a digital input on the PLC.

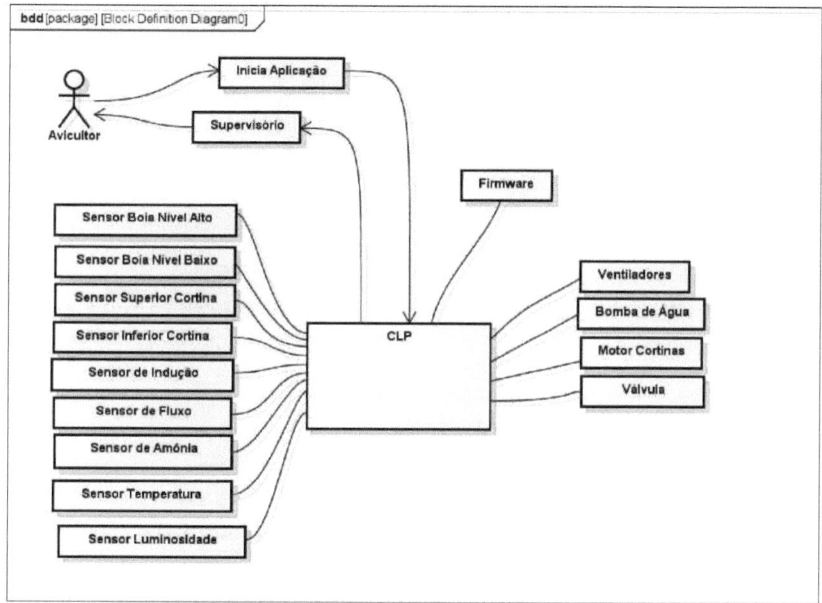

Diagram 6 - Block Diagram

Source: The author.

4.3 DEVELOPMENT OF SUPERVISORY *SOFTWARE*

The *WinCC Runtime Advanced* tool, which belongs to the TIA Portal package and is a specific tool for Siemens PLCs, was used to develop the supervisory system. WinCC is a *software* tool for all HMI applications, ranging from the most advanced operating solutions to the most advanced controllers.

from simple, basic panels to SCADA^ applications on PC-based multi-user systems (SIEMENS, between 2012 and 2017).

The supervisory system is structured with the components that are being managed, namely the tank and water consumption, feed consumption, activation and rearing period, and ammonia gas concentration, along with the activation of the motors and the valve. Figure 19 shows the values acquired by the sensors, conditioned by the *software* routines so that the farmer can see the information in an understandable way.

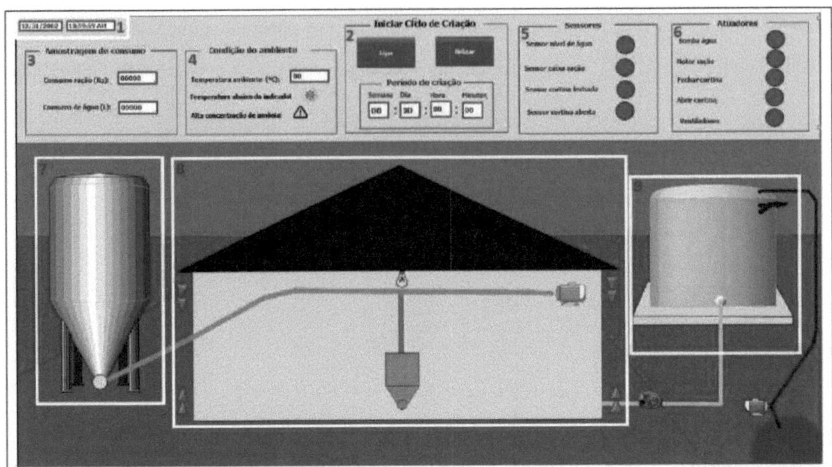

Figure 19 - Supervisory management

Source: The author

The supervisory image is subdivided into the following items:

1- Displays date and time information;

2- Iteration panel with the user, and with sampling of breeding time. This panel allows the poultry farmer to set the start and end of rearing;

3- Displays feed and water consumption information;

4- It displays information on the condition of the environment, showing the temperature value and displaying an alert when it is below ideal conditions, as well as informing you when there is a high concentration of ammonia;

[1] *supervisory control and data acquisition*

5- Shows sensor information: when they are activated they will be displayed in green, when deactivated in red;

6- Shows information about the actuators (motors) which are activated according to the sensor conditions. When deactivated, they will display red, and when activated, green;

7- Illustrates the silo where the feed is stored;

8- It shows a summarised view of the inside of the aviary where it displays some actuators, namely: light bulb, motor, fans and arrows to show the path of the curtain, these actuators will be displayed when activated;

9- Illustrates the water tank, which is made up of float-type level sensors to control the volume of water and activate/deactivate the water pump;

4.4 INTEGRATION BETWEEN THE EMBEDDED SYSTEM AND THE SUPERVISORY *SOFTWARE*

The integration between the *software* and the embedded system takes place through PROFINET communication, this standard is based on the Ethernet standard and represents communication with maximum transparency and open standard via TCP/IP, it contains network security and simultaneous real-time communication (SIMENS, between 2010 and 2017).

Integration was carried out by implementing *software* for the PLC with routines for checking ammonia gas, feed and hydration. The PLC was connected to sensors for checking ammonia gas, flow and an induction sensor to check the quantity of these inputs to be consumed by the birds. A supervisory system was also developed to display this information to the farmer.

4.5 RESOURCES AND BUDGETS

To develop the *hardware* with the programming routines for managing the variables and running tests, the teaching bench at UNOESC's automation laboratory was used.

Table 5 below shows the estimated budget for the components, with unit values, and the total value for the physical installation of the prototype in a poultry house.

Table 5 - Physical installation components and values

Component	**Quantity**	**Unit value**	**Total value**
CLP 1200 CPU 1214	1	2.880,00	2.880,00
LM18-3005PC Inductive Sensor	1	133,48	133,48
SM 1234 Analogue Module	1	1.450,00	1.450,00
Romak Gearmotor Q050 1/40 With Motor Brake 1/2 Cv	3	1.891,00	3.782,00
MQ-135 Gas Sensor	1	27,90	27,90
Water Flow Sensor l/2"YF-S201	1	34,90	34,90
Perforated Phenolite Board - 5x7cm	1	1,90	1,90
Materials for Mechanical Installations	1	1.200,00	1.200,00
Materials for Electrical Installations and Automation	1	2.225,00	2.225,00
Shipping	1	35,40	35,40
Total value of components			**11.774,38**

Source: the author

The money needed to develop the prototype, the equipment used and the electrical and mechanical installations will be provided by the poultry farmer. To carry out the tests, an S7-1200 PLC, an inductive sensor and power supplies were used to supply the PLC and the motors. These materials were provided by UNOESC.

Table 6 shows the amounts involved in purchasing the components to assemble the mock-up to demonstrate the prototype's operation.

Table 6 - Mock-up components and values

Component	**Quantity**	**Unit value**	**Total value**
Water level sensor	1	17,00	17,00
Miniature electric motor	4	17,80	71,20

Limit switch	5	2,50	12,50
Shipping	2	35,40	70,80
Materials for making a model	1	115,00	115,00
Model making	1	130,00	130,00
Total value of components			**416,50**

Source: the author

Table 7 shows which *software* resources were used to develop the prototype.

Table 7 - *Software* and Manufacturers

Software	**Manufacturers**	**Licence**
STEP 7 Professional V13 (TIA Portal V13)	Siemens	Professional
Astah SysML	Astah	Freeware
Bizagi Modeler	Bizagi	Freeware
Proteus	Labcenter	Trial
AutoCAD 2017	Autodesk	Free Trial

Source: the author

STEP 7 is Siemens' standard tool for automation systems, with which it is possible to develop the supervisory system and programme automation routines for PLCs. Its use is essential for developing the prototype *software* (SIEMENS, between 2010 and 2017).

The Astah SysML tool was used for UML modelling of *software, hardware,* installations and more. With this modelling, the prototype became simpler for the reader/user to understand. The use of this tool was necessary to be able to demonstrate the prototype's operating diagram (ASTAH, between 2010 and 2017).

Bizagi Modeler is a tool used to model processes, which makes it easier to understand how the *software* works and can be adjusted before implementation begins. This tool was essential for developing the variables to be managed; with the management logic programmed in the BPMN language, the development of the prototype became more effective (BIZAGI, between 2010 and 2017).

AutoCAD is used to develop architectural and structural prototypes, most commonly in civil engineering. The use of AutoCAD in the prototype was necessary to develop the floor plan of the aviary (AUTOCAD, between 2010 and 2017).

Proteus *software* is a tool that enables the electrical modelling and simulation of a circuit so that components can be added and removed by simulation before actually assembling them on a board (LTD, between 2010 and 2017).

The *software* described above, except for Bizagi Modeler, is available on the computers in UNOESC's automation and computer labs.

For the development of the miniaturised prototype integrating the components with the model, the values and components are shown in Table 8. The amounts used to develop the prototype were paid for by the author.

Table 8 - components and values

Components	**Quantity**	**Unit value**	**Total value**
MQ-135 Gas Sensor	1	27,90	27,90

Water Flow Sensor l/2"YF-S201	1	34,90	34,90
Perforated Phenolite Board - 5x7cm	3	1,90	5,70
Water level sensor	1	17,00	17,00
Miniature electric motor	4	17,80	71,20
Limit switch	5	2,50	12,50
Shipping	2	35,40	70,80
Materials for making a model	1	115,00	115,00
Model making	1	130,00	130,00
Total value			**485,00**

Source: the author

CHAPTER 5

TESTS CARRIED OUT

5.1 AMMONIA CONCENTRATION

As a precaution, the ammonia gas was replaced by butane gas to demonstrate the operation of the sensor and the system. In order to check for ammonia gas, the maximum concentration of the gas was set via *software* at 20 ppm. If this value is exceeded, information about the high concentration of ammonia in the environment will be displayed. To reduce the concentration of ammonia, a function was implemented to check the condition of the gas in the environment. When it is above the ideal, the *software* will provide correction methods, activating the fans, opening the curtains and issuing an alert on the supervisory system. Figure 20 shows an excerpt from the sensor's conditioning code so that it is activated and thus reports to the supervisor.

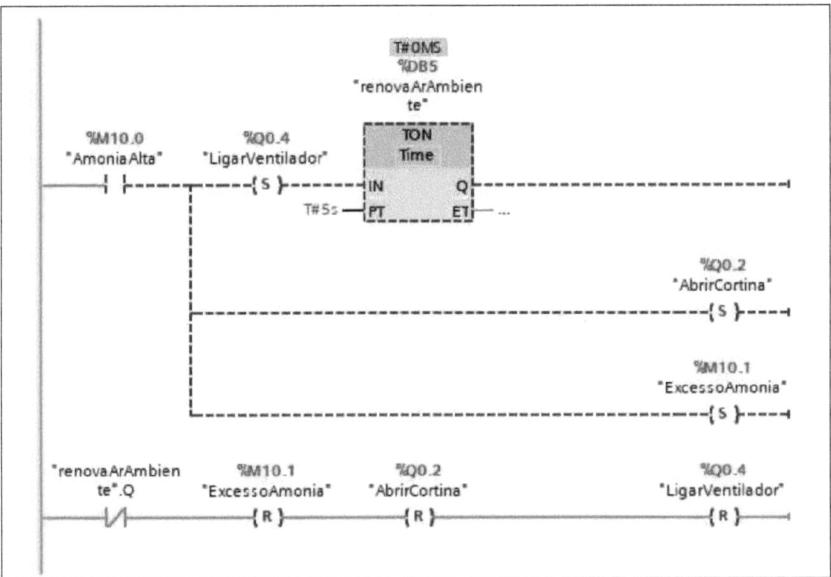

Figure 20 - Ammonia concentration check function

Source: The author

5.2 FEED CONSUMPTION

Feed consumption is calculated using a function that counts the number of revolutions the motor makes to reach one kilogram of feed. For this routine to be carried out accurately, a sensor is placed near the motor pulley responsible for transporting the feed. This pulley will have a metal plate, so every time it passes the inductive sensor it will be stored in an internal counter, making it possible to calibrate the sensor. Figure 21 shows a section of the code that executes the feed consumption routine.

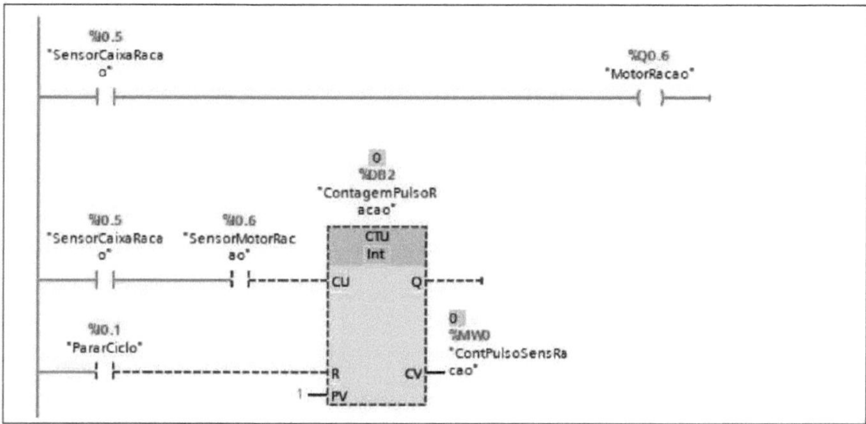

Figure 21 - Feed consumption verification function

Source: The author

5.3 WATER CONSUMPTION

To check water consumption, the flow sensor was used, which, when water passes through, turns an internal propeller that has a magnet attached to it, making it possible to capture data emitted by the sensor. The amount of litres of water consumed is counted using the data collected from the sensor. The code used to configure this routine is shown in Figure 22.

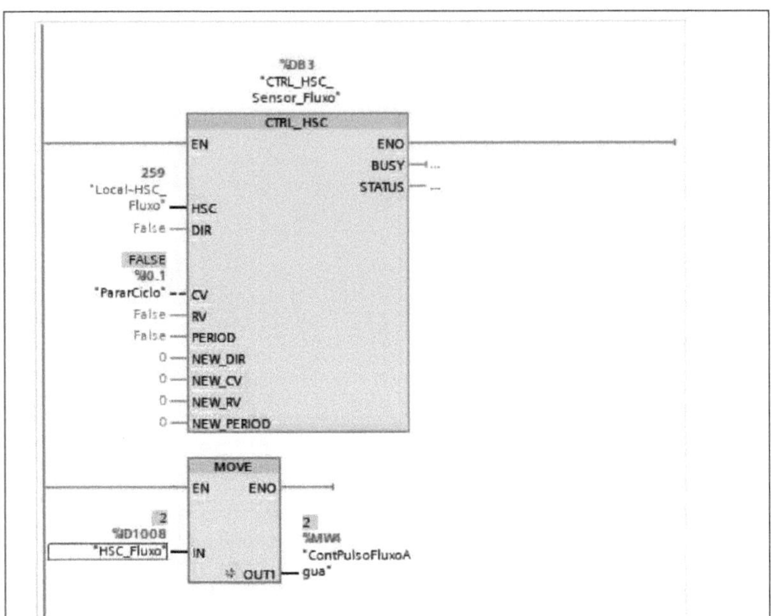

Figure 22 - Function for checking water consumption
Source: The author

5.4 LIGHT SUPPLY AT NIGHT

The routine for supplying electricity at night is based on the information provided by the slaughterhouse with which the poultry farmer is associated. It uses a light sensor connected to a transistor, which in turn is connected to a relay, so that when the light decreases, it sends a signal to the PLC, which executes the programmed routine for supplying light in the poultry's rearing environment. The partial code for this routine is shown in Figure 23.

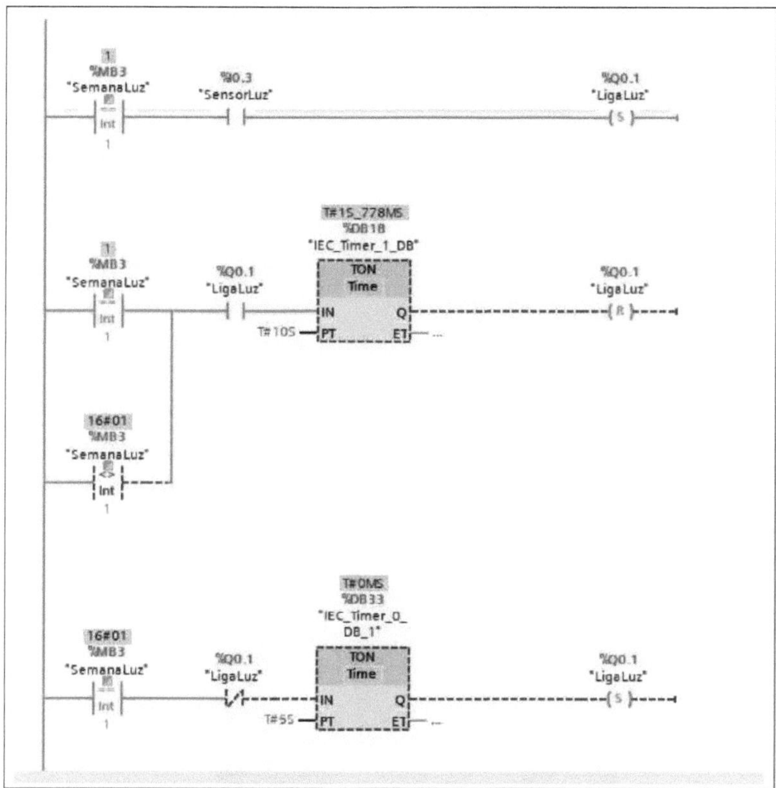

Figure 23 - Bench test
Source: The author.

5.5 *HARDWARE* AND *SOFTWARE* INTEGRATION

After developing and optimising the routines, the integration of the sensors began, and tests were carried out with the programmed routines being executed in the PLC. All the tests were carried out in the laboratory using prototyping. Figure 24 illustrates the sensors and actuators integrated into the PLC, as well as the supervisory system that shows the condition of the poultry house environment, and the actions being applied when the environment is not suitable for the farmer.

The tests were carried out using two sources, one set at 24 V to switch on the PLC, the flow and induction sensors, and the motor to simulate the curtain, and the other at 5 V to switch on the gas sensor that will detect ammonia. The PLC is connected to the computer via a network cable for communication, thus sending data acquired by the PLC to the supervisor. Figure 24 shows a bench test with some of the components connected to the PLC.

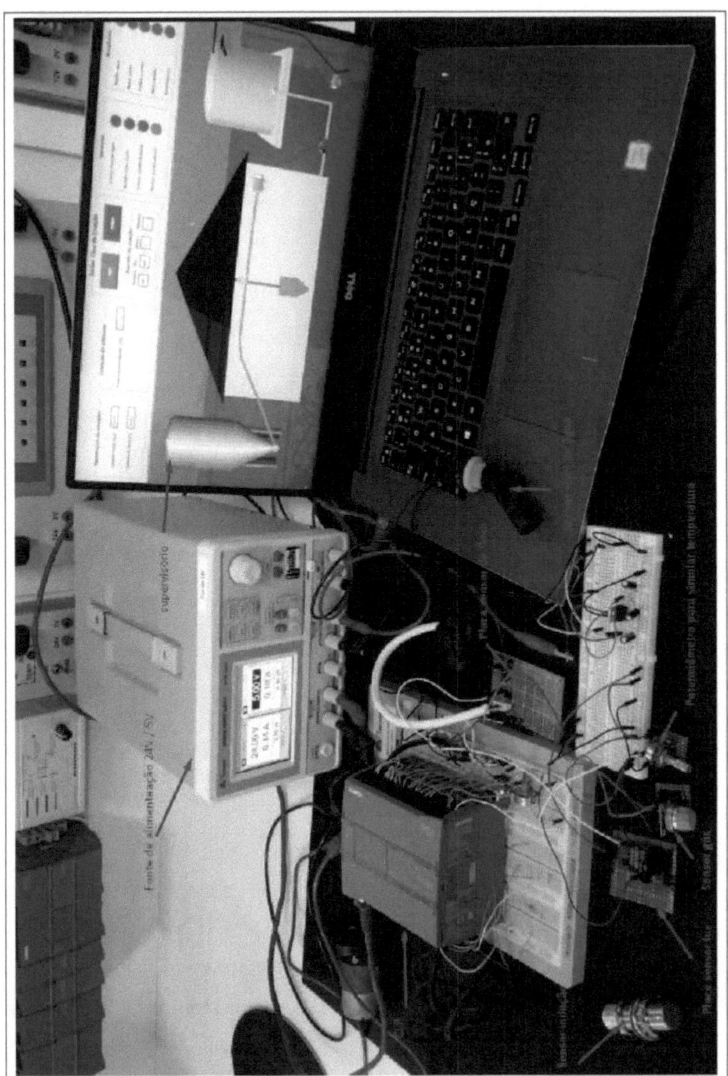

Figure 24 - Bench test

Source: The author.

The prototype is configured to start monitoring when the poultry farmer activates an external switch connected to digital input 6 of the PLC, or accesses it via the supervisory control centre and makes the command to start breeding, this command being set via a button on the supervisory control centre screen. Once creation has started, the sensors will begin their checks. In the water tank, the float-type sensor near the edge of the tank will be responsible for activating/deactivating the water pump. The sensor at the bottom is designed to close the valve when the tank has a low water level, and open it when the water level is above the sensor.

The sensor that will send the signal to the PLC to calculate the amount of feed consumed by the birds will be activated when the feed box sensor (already installed) deactivates, thereby activating the motor that takes the feed to the boxes, enabling the induction sensor to act and make the appropriate calculation and inform the supervisory system. The water flow sensor will be constantly active, so when water flows through the pipe, it will take a reading and, using the routine programmed for it, send the amount of litres of water consumed by the birds to the supervisory system. The ammonia gas sensor will constantly check the environment. When the ammonia level exceeds the recommended level (20 ppm), the PLC will detect the signal sent by the sensor and send a signal to the motors that will lower the curtains and switch on the fans for 20 seconds.

CHAPTER 6

TESTS AND RESULTS

This section shows the tests carried out to check that the prototype works. Figure 25 illustrates how the routine for checking the water in the tank works: when the level sensor is deactivated, the supervisor is informed that the water pump actuator is activated and the tank is informed that the sensor is deactivated.

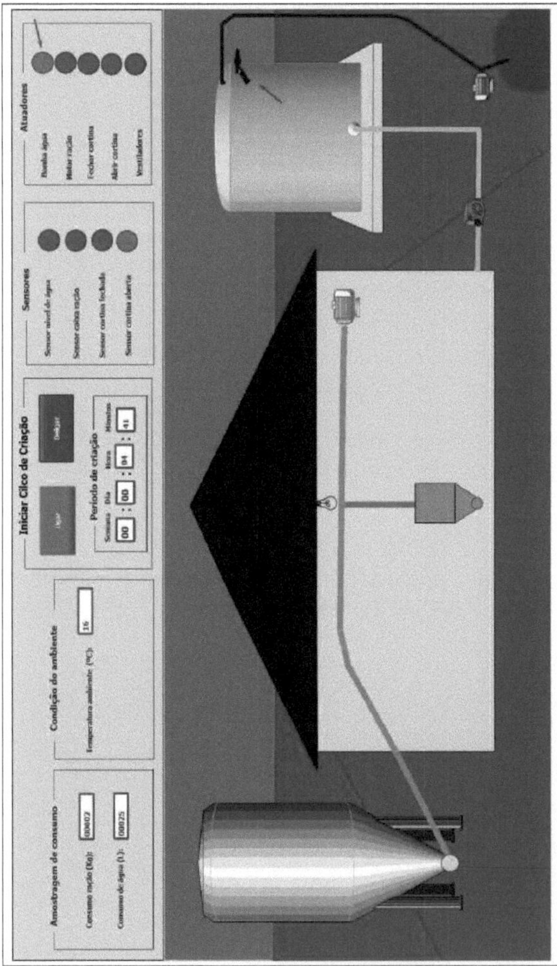

Figura 25 - Routine water test

Source: The author.

Figure 26 shows the operation of the lighting routine, which checks whether the sensor is activated and indicates a lack of light, thus informing the supervisory system and activating the electric light in the poultry's rearing environment.

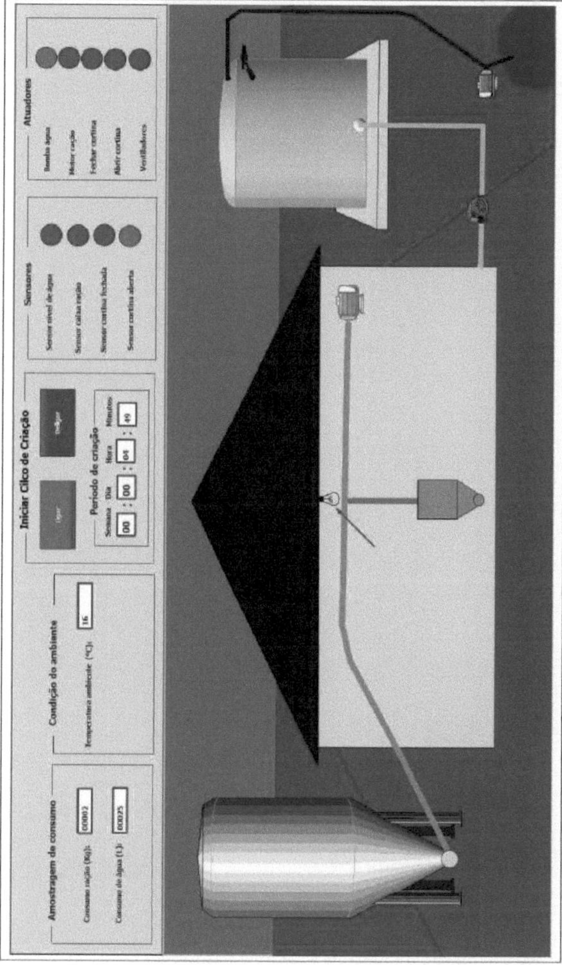

Figura 26 - Test light routine

Source: The author.

Figure 27 shows the treatment coded to inform the creator of the situation of the environment via the supervisory system, applying the correction and activating the fans.

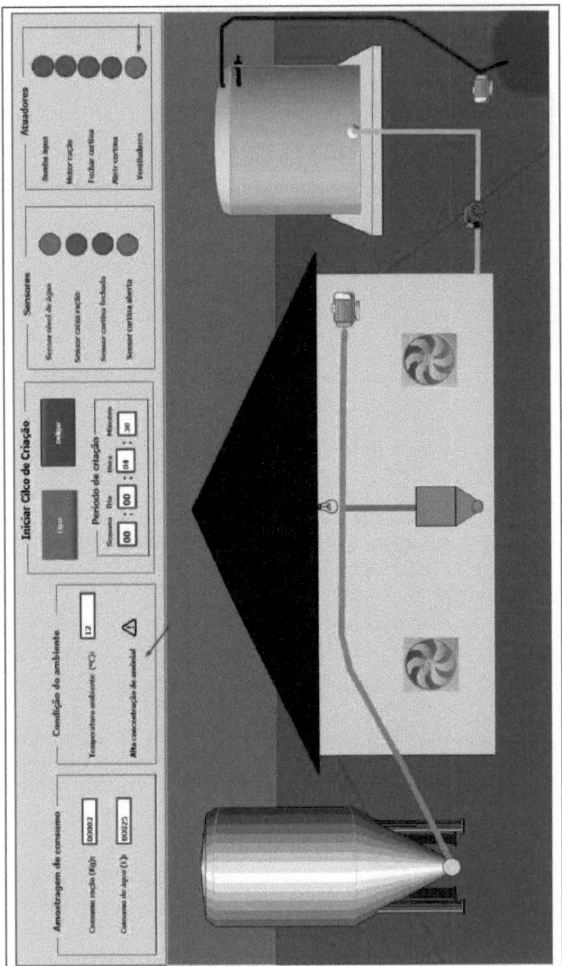

Figure 27 - Routine ammonia test
Source: The author.

Figure 28 shows the prototype designed to validate the operation of the sensors and actuators to keep the birds' rearing environment stable.

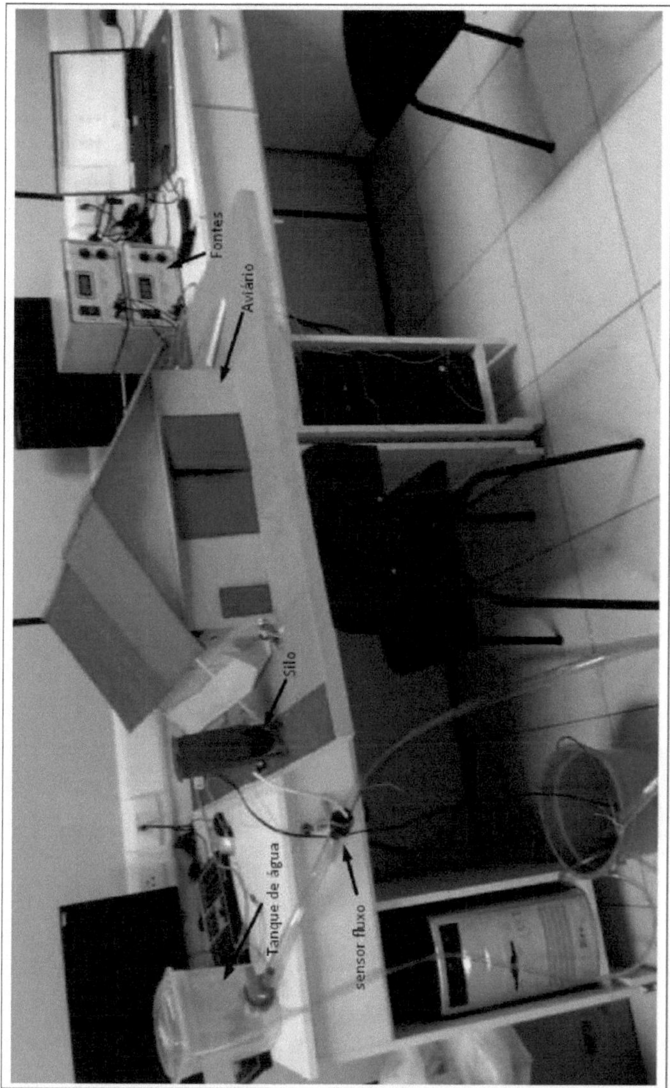

Figure 28 - Prototype

Source: The author.

Figure 29 shows the connection between the sensors and the motor, which are responsible for opening and closing the aviary curtains as required by the internal environment.

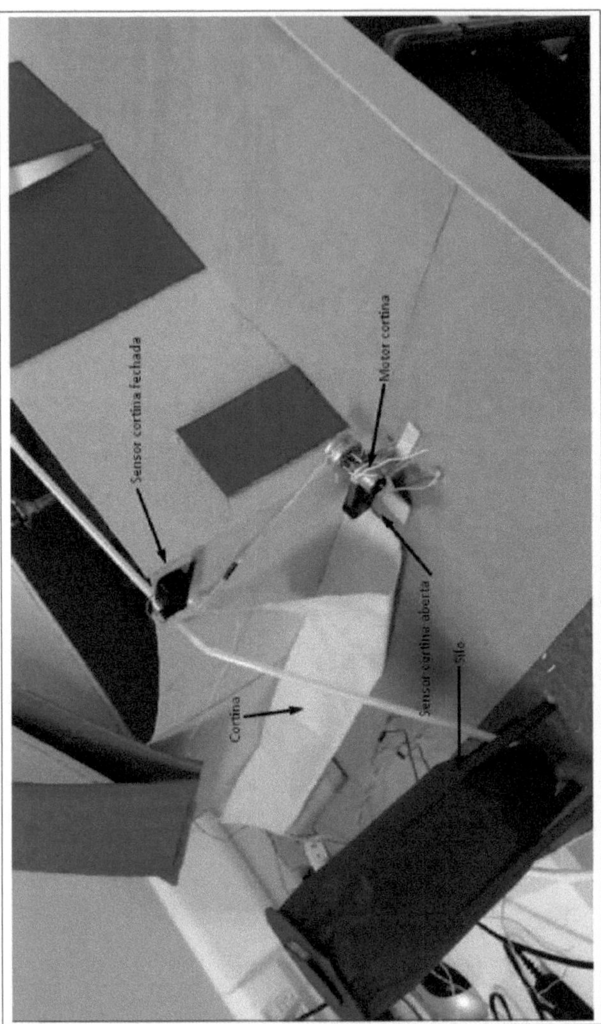

Figure 29 - Curtain sensors

Source: The author.

Figure 30 shows a preview of the internal environment of the aviary, where you can see the fan, the light and the feed bin, together with the line of feeders, as well as the motor that is responsible for moving the feed from the silo to the feed bin and the sensor that is used to monitor feed consumption.

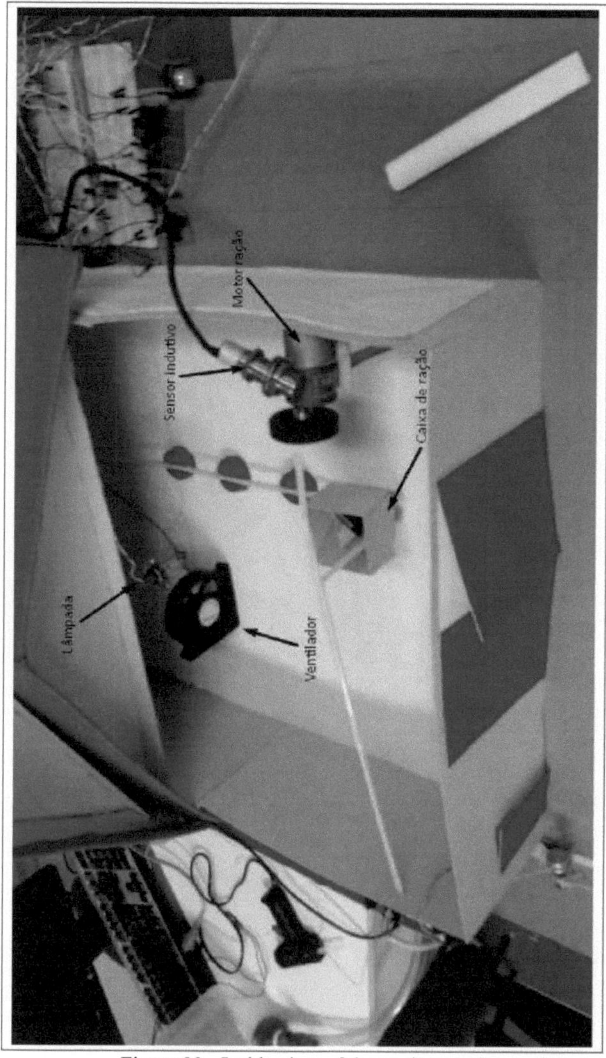

Figure 30 - Inside view of the poultry rearing environment

Source: The author.

The boards with the sensors integrated into the PLC can be seen in Figure 31. A potentiometer is being used to simulate and demonstrate the temperature variation.

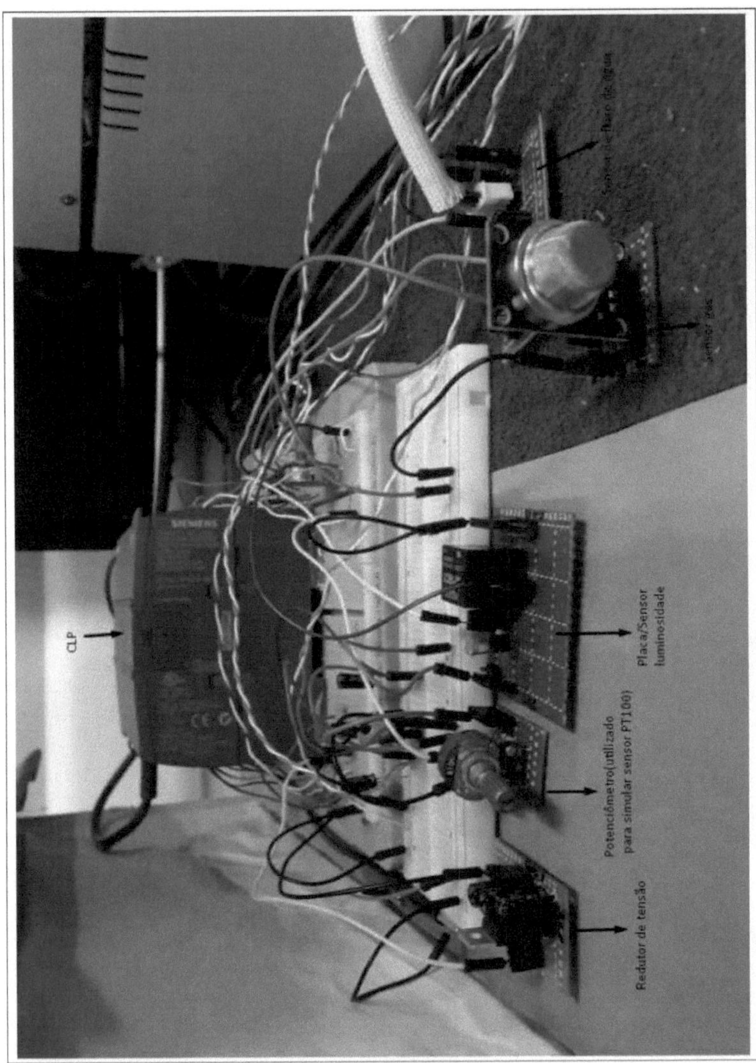

Figure 31 - Integration of sensor boards with PLC

Source: The author.

CHAPTER 7

FINAL CONSIDERATIONS

It can be concluded that this prototype has been realised with the aim of developing an autonomous application for managing variables that directly influence the poultry rearing sector, the correct management of these environmental variables preventing some of the problems encountered during the rearing process. To this end, a study was carried out on the appropriate environment to generate thermal comfort for poultry, through a bibliographical survey, which includes scientific articles related to this area.

An important point to consider when implementing the project *on site is* that the poultry house must be semi-automatic, where the feed supply system from the feed box to the bird feeders must already be automated. This way, it is ready for the alterations to be made and the sensor to be added so that feed consumption can be measured.

The prototype has sensor verification routines, which send electrical signals to the PLC and are translated by the algorithm applied to it, thus applying ventilation, lighting, water and feed control and ammonia gas concentration verification tasks, followed by corrections to provide an environment that provides ideal thermal comfort or close to it. The prototype was developed and tested in UNOESC's automation laboratory.

Some of the most significant challenges encountered in the development of this prototype were due to the choice of a water flow sensor that is not suitable for the PLC, requiring a good deal of time studying and searching for information to condition the signal, which is 5 V to more than 10 V, and to condition the *software* to interpret the signal on its digital port.

In the general context of the prototype, the integration of the curricular internship projects I and II were adapted to the Final Course Work. As for the calibration of the algorithm, the code and routines were optimised so that the system works autonomously, in order to provide an environment that provides thermal comfort for the birds. The boards, to which the sensors were connected, were designed using islanded boards, making it easier to replace and add components if necessary. Once the boards had been made and the sensors tested, they were integrated into the prototype. After integration, the supervisory system was developed to display the information collected by the sensors and show the actuators acting to maintain the quality of the environment. Finally, the workings of the prototype were demonstrated to the members of the judging panel.

CHAPTER 8

FUTURE WORK

With the knowledge acquired in the development of the prototype, it is possible to define some future objectives that will contribute to the current prototype:

- Integrate internal PLC databases with external databases;
- Develop a hybrid application that poultry farmers can access from anywhere and from any device with an internet connection;
- Integrate camera monitoring;
- Develop reports with information to visualise the birds' development;
- Display the birds' development on a graph, comparing it with information from their breeding history.

REFERENCES

5PIGR0UP. *Broiler feeding system Layout.* between 2010 and 2017.

ASTAH. **Astah SysML Tutorial.** between 2010 and 2017. Available at: <http://astah.net/tutorials/ sysml/intro>. Accessed on: 09.04.2018.

AUTOCAD. **Autocad.** between 2010 and 2017. Available at: <https://www.autodesk.com.br/ products/autocad/overview>. Accessed on: 09.03.2018.

Poultry, Portal Suínos e. **Steps in broiler handling.** 2016. Available at: <http://www.aviculturaindustrial.com.br/imprensa/etapas-do-manejo-de-frango-de-corte/ 20130307-090133-h028>. Accessed on: 08.06.2018.

AVILA, VS de et al. Production and management of broiler chickens. **Concórdia: EMBRAPA-CNPSA,** 1992.

BASSI, Levino José et al. Basic recommendations for the management of colonial broilers. EMBRAPA, 2006.

BIGSAL, Animal Nutrition. Producer's booklet 1: Feed management for broiler chickens, between 2006 and 2016. Available at: <http://www.bigsal.com.br/ manejo-alimentar-de-frangos-de-corte.php>. Accessed on: 08.02.2018.

BIZAGI. **Main benefits of Bizagi Modeler.** between 2010 and 2017. Available at: <http://www.bizagi.com/pt/produtos/bpm-suite/modeler>. Accessed on: 09.03.2018.

CARE. Broiler chickens. 2014.

COOB-VANTRESS. Broiler handling manual. 2012. Accessed on: 29.05.2018.

. Supplement: Performance and nutrition for broilers. 2012. Available at: <http: //www.cobb-vantress.com/languages/guideiiles/793al6cc-5812-4030-9436-le5dal77064f_pt. pdf>. Accessed on: 08.02.2018.

FERREIRA, João Coutinho. **Removal of ammonia generated on poultry farms and its use in fuel cells and as fertiliser.** Thesis (Doctorate) - University of São Paulo, 2010.

Gl. Sc slaughters 1 billion chickens a year and is the country's 2° largest exporter. 2015. Available at: <http://gl.globo.com/sc/santa-catarina/campo-e-negocios/noticia/2015/10/ sc-abate- l-bilhao-de-frangos-por-ano-e-e-o-2-major-exportador-do-pais.html>. Accessed on: 08.06.2018.

GEORGINI, Marcelo. **Applied Automation Description and Implementation of Sequential Systems in PLCs.** 9th ed. São Paulo: Érica Ltda., 2012.

INSTITUTE, Digital. **Water Flow Sensor 1/2 - YF-S201.** between 2010 and 2017. Available at: <http://www.institutodigital.com.br/ pd-24b6c2-sensor-de-fluxo-e-vazao-de-agua-l-2-yf-s201.html>. Accessed on: 05.04.2018.

LTD, Labcenter Electronics. **Proteus Design Suite.** between 2010 and 2017. Available at: <https://www.labcenter.com/>. Accessed on: 09.04.2018.

MAPA, Ministry of Agriculture, Livestock and Supply. Poultry, between 2008 and 2016. Available at: <http://www.agricultura.gov.br/animal/especies/aves>. Accessed on: 30.03.2018.

MIRAGLIOTTA, MIWA YAMAMOTO. Evaluation of internal environmental conditions in two commercial broiler houses with different ventilation and stocking density. **Campinas: Faculty of Agricultural Engineering, State University of Campinas,** 2005.

MOTA, Allan. **Light Sensor - Learning to use the LDR.** 2015. Available at: <http://blog.vidadesilicio.com.br/arduino/basico/sensor-de-luz-ldr/>. Accessed on: 26.06.2018.

NOVUS. **PtlOO. thermoresistances** between 2010 and 2016. Available at: <http://www.nuvus. com.br/downloads/Archivos/folheto_ptlOO.pdf>. Accessed on: 29.04.2018.

OLIVEIRA, Paulo Armando Victoria de; MONTEIRO, Alessandra Nardina Trícia Rigo. Ammonia emissions in broiler production. In: IN: FACTA CONFERENCE, CAMPINAS, 2013. ANNALS... CAMPINAS: FACTA, 2013. 1 CD-ROM. **Embrapa Swine and Poultry-Article in conference proceedings (ALICE).** [S.I.], 2013.

OMG, OBJECT MANAGEMENT GROUP. **OMG Systems Modelling Language.** [S.I.], 2010. Manual. Available at: <http://www.vidadesilicio.com.br/ sensor-de-gas-mq-135-amonia-oxido-nitrico-alcool-benzeno-dioxido-de-carbono-e-fumaça. html>.

OWADA, Adriana N et al. Estimation of broiler welfare as a function of ammonia concentration and light level in the production house. **Agricultural Engineering,** Brazilian Association of Agricultural Engineering, 2007.

PÁDUA, Elisabete Matallo Marchesini de. **Metodologia da pesquisa: abordagem teórico-prática.** Campinas: Papirus, 2004.

PAREDE, Ismael Moura; GOMES, Luiz Eduardo Lemes. Electronics: Industrial Automation, v. 6, 2011. Available at: <http://eletro.gl2.br/arquivos/materiais/eletronica6.pdf>. Accessed on: 18.04.2018.

PORTA, Leonardo Dalla. **Water flow sensor for arduino 1-30 L/min.** 2016. Available at: <http://blog.usinainfo.com.br/sensor-de-fluxo-de-agua-para-arduino-l-30-lmin/>. Accessed on: 05.06.2018.

PRUDENTE, Francesco. **PLC Industrial Automation: Theory and Applications.** 2nd ed. Rio de Janeiro: gen, 2013.

REISDORFER, Rodrigo. **Automation of a grinding process for animal feed.** 2011. TCC (Electrical

Engineering Course Final Paper), UNOESC (UNIVERSIDADE DO OESTE DE SANTA CATARINA), Joaçaba, Brazil.

ROSS. Breeding stock management manual. 2013. Available at: <http://www.aviagen.com>. Accessed on: 08.10.2016.

SENSE SENSORS AND INSTRUMENTS. **Inductive sensors instruction manual.** [S.l.J, 2002. Manual.

SIEMENS. **Fully Integrated Automation Portal,** between 2000 and 2015. Available at: <http://w3.siemens.com.br/automation/br/pt/tia-portal/portal-automacao-tia/pages/default. aspx>. Accessed on: 01.05.2018.

. **STEP 7 Professional.** between 2005 and 2015. Available at: <http://w3.siemens.com/ mcms/simatic-controller-software/en/step7/step7-professional/Pages/Default.aspx>. Accessed on: 01.05.2018.

. **TIA Portal - The integrated engineering framework that redefines engineering,** between 2005 and 2016. Available at: <http://w3.siemens.com.br/automation/br/pt/tia-portal/Pages/ Default.aspx?ismobile=true>. Accessed on: 01.05.2018.

. **SIMATIC STEP 7 V5.x the proven engineering system.** between 2010 and 2017. Available at: <http://w3.siemens.com/mcms/simatic-http://w3.siemens.com/mcms/simatic-controller-software/en/step7/pages/default.aspx>. Accessed on: 09.03.2018.

. **SIMATIC WinCC (TIA Portal) - Runtime Software,** between 2012 and 2017. Accessed on: 11.05.2018.

SILÍCIO, Vida de. **MQ-135 Gas Sensor - Ammonia, Nitric Oxide, Alcohol, Benzene, Carbon Dioxide and Smoke,** between 2010 and 2017. Available at: <http://www.vidadesilicio.com.br/ sensor-de-gas-mq-135-amonia-oxido-nitrico-alcool-benzeno-dioxido-de-carbono-e-fumaca. html>. Accessed on: 05.06.2018.

SILVEIRA, Cristiano Bertulucci. **Inductive Sensor: What it is and how it works.** 2016. Available at: <https://www.citisystems.com.br/sensor-indutivo/>. Accessed on: 05.06.2018.

. **How the LADDER Language Works.** between 2005 and 2018. Available at: <https://www.citisystems.com.br/linguagem-ladder>. Accessed on: 18.04.2018.

SIMENS. **Integrated PROFINET communication** between 2010 and 2017. Available at: <http://w3.siemens.com.br/automation/br/pt/automacao-e-controle/automacao-industrial/ simatic-plc/s7-cm/SIMATIC-S7- 1200/comunicacao/Pages/Comunicacao.aspx>. Accessed on: 08.06.2018.

SIRICHOTE, Wichit. **Handheld PtlOO Thermometer.** between 2010 and 2016. Available at: <http://www.kswichit.corn/ptlOOhandheld/PtlOOhandheld.htm>. Accessed on: 25.02.2018.

SUNROM TECHNOLOGIES. **Light Dependent Resistor - LDR.** [S.I.], 2008. Datasheet.

THOMAZINI, Daniel; ALBUQUERQUE, Pedro Urbano Braga de. **Sensores Industriais - Fundamentos e Aplicações.** 8th ed. São Paulo: Editora Érica Ltda., 2012.

USINAINFO. **Water Level Sensor with Horizontal Float,** between 2010 e 2016. Available at: <http://www.usinainfo.com.br/sensores-para-arduino/ sensor-de-nivel-de-agua-com-boia-horizontal-2580.html>. Accessed on: 25.02.2018.

WAVESHARE. **MQ-135 Gas Sensor User Manual.** [S.I.], Between 2000 and 2012. Manual.

WORDPRESS. **Contact diagram (LADDER symbology).** between 2005 and 2014. Available at: <https://automacaodiaria.files.wordpress.com/2014/08/slide_12.jpg>. Accessed on: 09.06.2018.

ZANATA, Rodrigo Antonio. **Analysing ammonia control in poultry farms.** 2007. Monograph

(Specialisation in Occupational Safety Engineering), UNESC (UNIVERSIDADE DO EXTREMO SUL CATARINENSE), Criciúma, Brazil.

ZANATTA, Jean Luiz. **System for monitoring ammonia gas in poultry houses.**
2014. TCC (Final Coursework in Computer Engineering), UNOESC (UNIVERSIDADE DO OESTE DE SANTA CATARINA), Joaçaba, Brazil.

Printed by Books on Demand GmbH, Norderstedt / Germany